PROPERTY OF
MORTON GROVE
SCHOOL DISTRICT 70

Around the Moon 1, 2, 3

A Space Counting Book

by Tracey E. Dils

AMICUS READERS 1 AMICUS INK

amicus readers

Say Hello to Amicus Readers.

You'll find our helpful dog, Amicus, chasing a ball—to let you know the reading level of a book.

1 Learn to Read
Frequent repetition, high frequency words, and close photo-text matches introduce familiar topics and provide ample support for brand new readers.

2 Read Independently
Some repetition is mixed with varied sentence structures and a select amount of new vocabulary words are introduced with text and photo support.

3 Read to Know More
Interesting facts and engaging art and photos give fluent readers fun books both for reading practice and to learn about new topics.

Amicus Readers and Amicus Ink are imprints of Amicus
P.O. Box 1329, Mankato, MN 56002
www.amicuspublishing.us

Copyright © 2016. International copyright reserved in all countries. No part of this book may be reproduced in any form without written permission from the publisher.

Library of Congress Cataloging-in-Publication Data
Dils, Tracey E., author.
 Around the moon 1, 2, 3 : a space counting book / by Tracey E. Dils.
   pages cm. -- (1, 2, 3... count with me)
 Summary: "Introduces the moon, planets, and other objects in space, while teaching the concept of counting to ten"-- Provided by publisher.
 Audience: Grades K to 3.
 ISBN 978-1-60753-714-4 (library binding)
 ISBN 978-1-60753-818-9 (ebook)
 ISBN 978-1-68152-000-1 (paperback)
 1. Counting--Juvenile literature. 2. Moon--Juvenile literature. 3. Solar system--Juvenile literature. I. Title.
 QA113.D5525 2015
 513.2'11--dc23
                       2014045267

Photo Credits: Shutterstock Images, cover (background), cover (top left), cover (bottom left), cover (right), 1, 4-5, 5, 6, 14-15, 14 (top right), 14 (top left), 14 (bottom right), 14 (bottom left), 20-21, 22-23, 24 (top left), 24 (top right, top), 24 (top right, middle), 24 (bottom right); Russell Croman/NASA, 3; NASA, 9; NASA Planetary Photojournal, 10-11, 18-19, 24 (bottom left); Marc Ward/Shutterstock Images, 12-13; JPL-Caltech/MSSS/NASA, 13 (top left), 13 (top right), 13 (middle), 13 (bottom left), 13 (bottom right); Andrea Crisante/Shutterstock Images, 15 (top), 24 (top right, bottom); Yurji Omelchenko/Shutterstock Images, 15 (bottom); Igor Kovalchuk/iStock/Thinkstock, 16-17; SAO/CXC/STScI/NASA, 20 (top left); JPL-Caltech/NASA, 20 (top right); Kate Su/JPL-Caltech/NASA, 20 (bottom left); J. Hester/ASU/ESA/NASA, 20 (bottom right); Subaru Telescope (NAOJ)/Hubble Legacy Archive/NASA, 21 (middle left); Josepy Brimacombe/NASA, 21 (top); K.Getman, E.Feigelson, M.Kuhn & the MYSHX team/PSU/CXC/JPL-Caltech/NASA, 21 (bottom left); Hubble/ESA/NASA, 21 (middle right); Brian Davis/NASA, 21 (bottom right)

Produced for Amicus by The Peterson Publishing Company and Red Line Editorial.

Editor Jenna Gleisner
Designer Craig Hinton

Printed in Malaysia
HC 10 9 8 7 6 5 4 3 2 1
PB 10 9 8 7 6 5 4 3 2 1

Outer space is full of fun things to count. Let's count them!

1

One moon shines at night. The moon circles the earth.

2

Two rockets shoot up into the sky. Rockets take astronauts into space.

3

Three astronauts work in space. They wear suits and helmets.

9

4

Four large moons circle Jupiter. Io is yellow.

5

Five pictures of Mars are sent to Earth. A Mars rover drives on Mars and takes pictures.

13

6

Six satellites circle the earth. They take pictures of outer space.

7

Seven stars make the Big Dipper. It is also called the Big Bear.

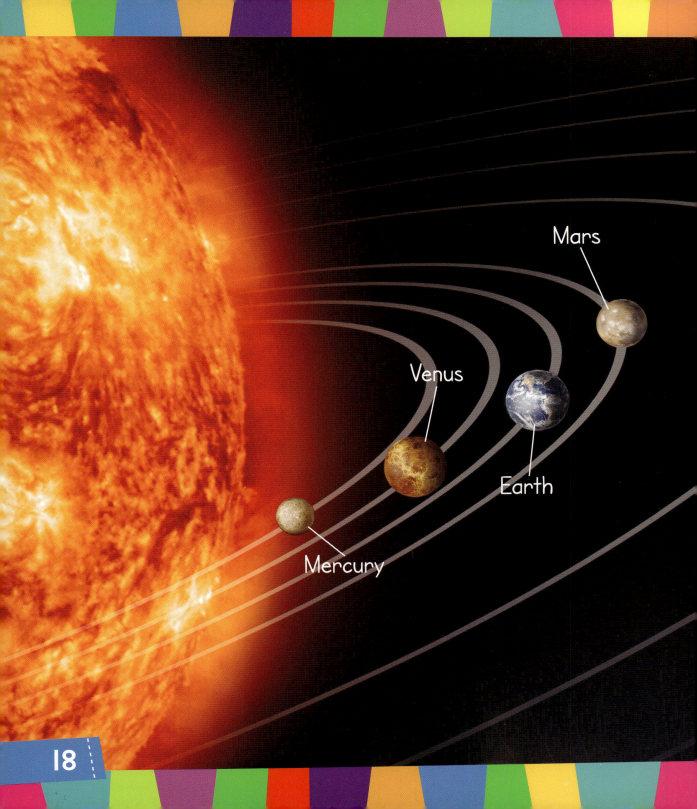

Eight planets are in our solar system. Saturn has big rings.

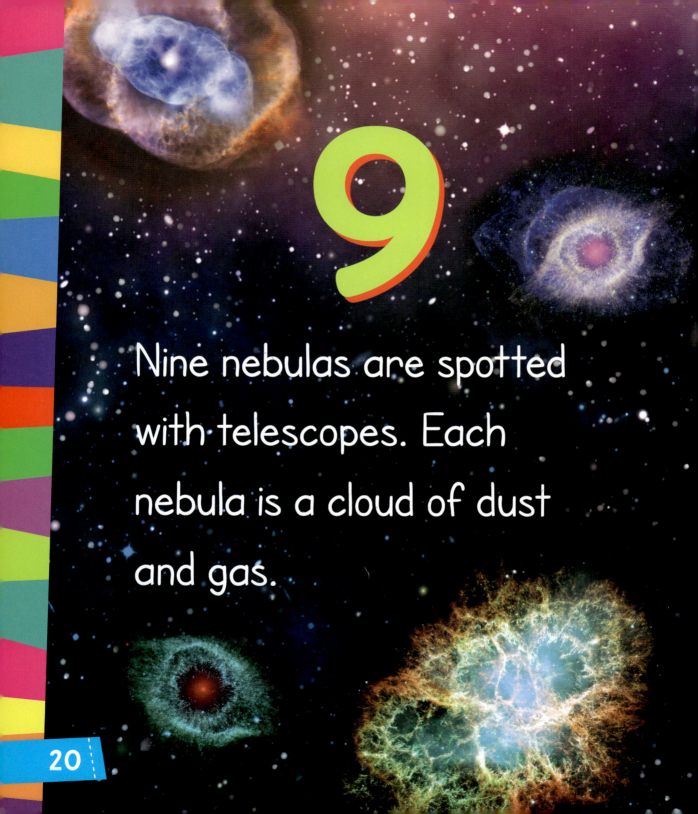

9

Nine nebulas are spotted with telescopes. Each nebula is a cloud of dust and gas.

10

Ten meteors light up the night sky. What else can you count?

Count Again

Count the number of objects in each box.

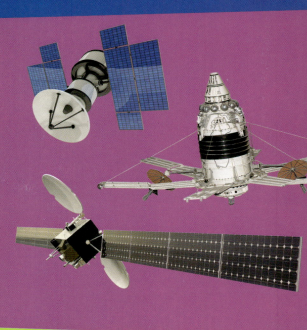